The Mathematics from the Mayas

And Japanese, Nahuatl and Egyptian are similar languages."

Written and author by Erik De La Torre Stahl(Erik Stahl)
Copyrights 1986

In the beginning the human being tried

to have communication each other.

There was no language, only sounds and

shouts between them. Around them there

was nature, sounds

of birds and the sea. What was the first

word pronounced for the human being.

Is a question that sometimes we ask

ourselves. By happiness, pain, suffering,

crying or laughing, who knows what

makes for first time pronounce a word.

The music is part of the nature, there is

no nature with out music and there is no

music with out nature.

All the musical instruments are in

harmony with the nature. The song of

the bird can stop even the lion in battle.

Some natives in South Africa speak with

a mix of whistles and words. Probably in

the beginning the human being started

talking with whistles or other

sounds. By the time those whistles and

sounds made words, those words made

drawings.

With simple drawings the human started

making communication, like the

Egyptians, the Nahoas from Mexico and

the Orientals.

The Egyptians started drawing semiotic

Symbols like legs that means walk, or an

eye that means watch something. But

those symbols were followed with an

alphabet. In the new world, or rather say

Continent of America, The Mayans

created syllabus by drawing

faces, parts of the human body. Even

they invented the zero, some of the faces

means numbers of days. The next

illustration explain how to know the

exactly date of the stela.

For me took me 1 year to discovered

in 1984 the way to read the date of the

Mayan's Stelas.

Here an example

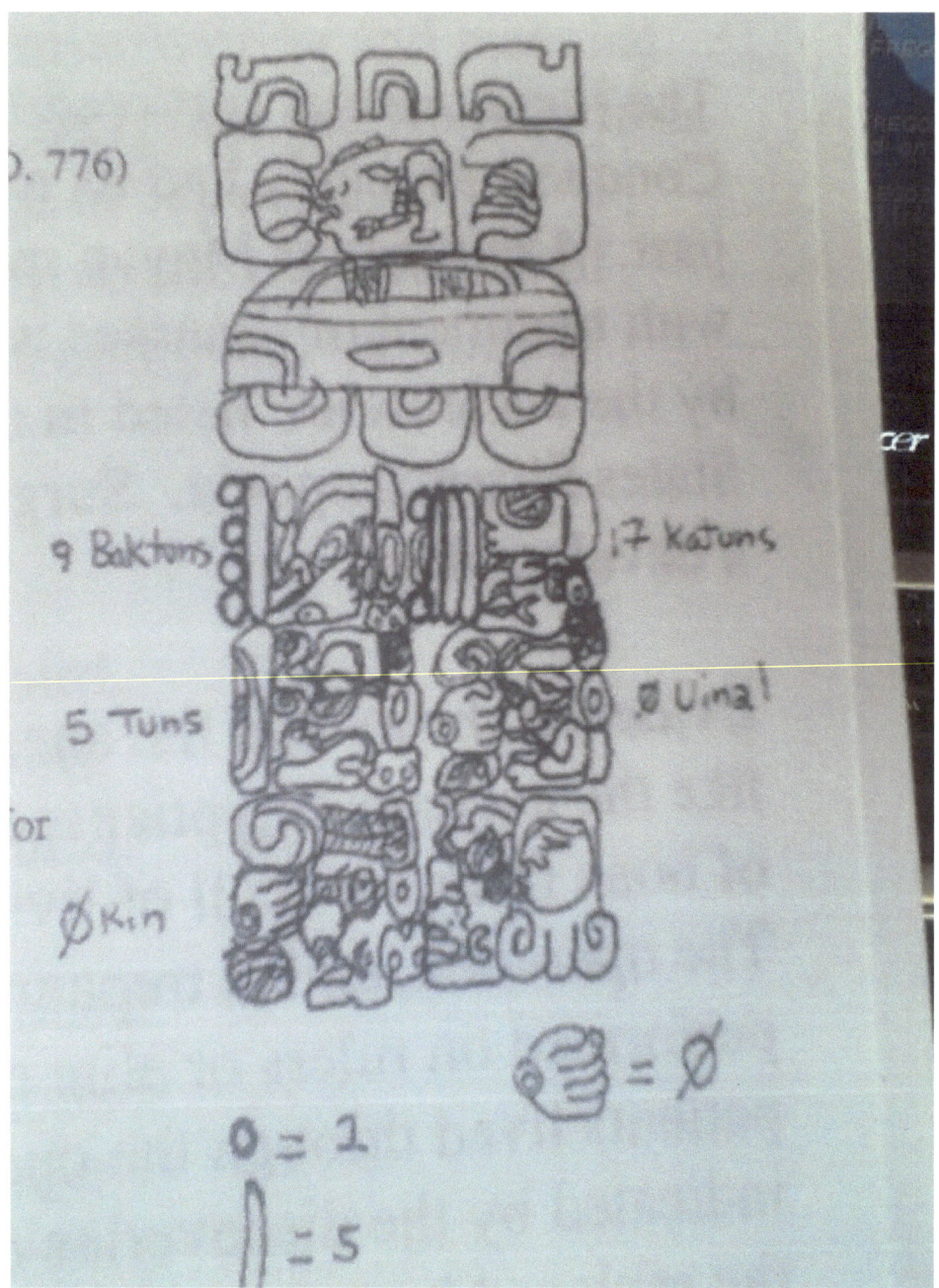

(D. 776)

9 Baktuns ¡7 Katuns

5 Tuns Ø Uinal

or

Ø Kin

🖐 = Ø

O = 1

𝇋 = 5

Stella 1 Quirigua 9.17.5.0.0 (A.D. 776)

In Tikal, Guatemala.

9 Baktun x 144,000 Days-   1,296,000 days

17 Katun x   7,200 Days-   122,400 days

5   tun x   360 Days   1,800 days

1,420,200/365=3,890 years ago

3,890 -

114 days means , days no-work for Mayas

_____

3,776

3,776 –

3,000

  776 A.D

In Maya's numbers, the symbol "o"

Means "One day"

The symbol "__" means number 5 days

The MAYAs' Mathematics

Here an example how the Mathematics from the Mayas was faster to resolve than our European system .

125 X 235= 29375 MAYA'S MATH

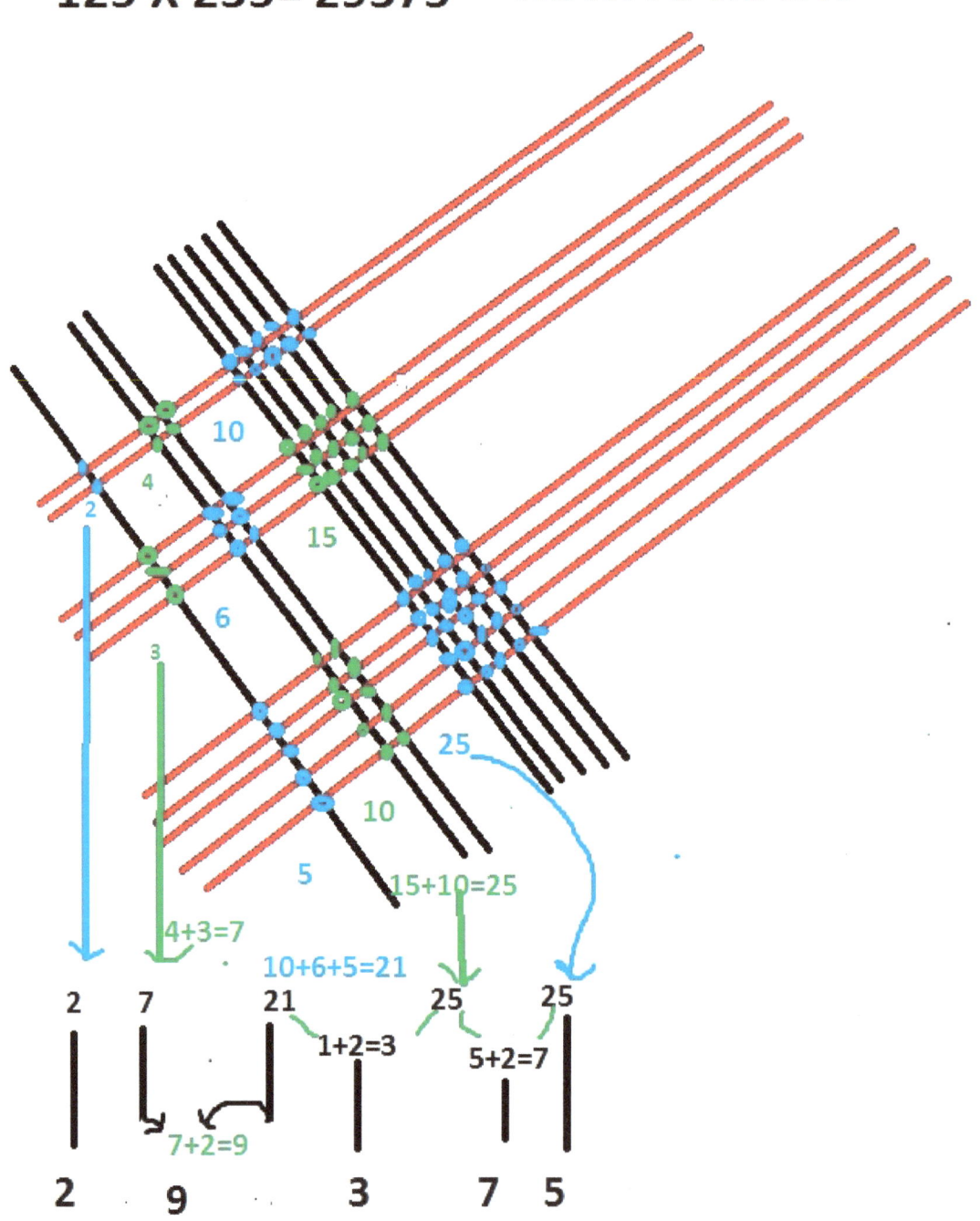

MAYA'S MATH

$5 \times 360 = 1800$

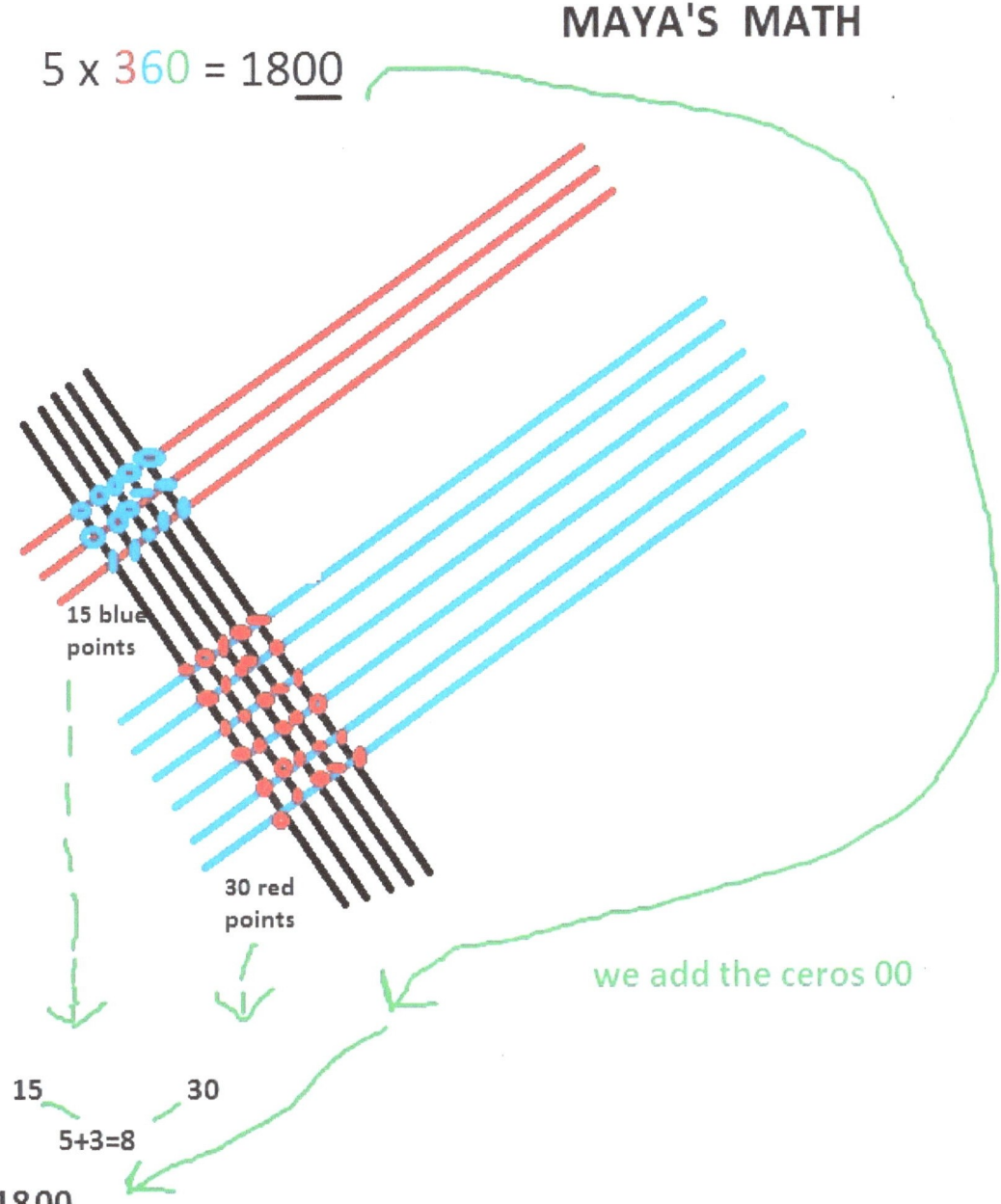

15 blue points

30 red points

we add the ceros 00

15 30

5+3=8

1800

The Aztecs, Mayans and the Egyptians

in many ways are similar. Like the

pyramids, many words, even in medicine

did have medical literature involving

medicinal

plants.

The literature that survived the Spanish

Conquest probably had its roots in the

ancientpast. A number of Mayan medial

texts deal with treatment of illnesses

with herbs used by the ancients are listed

in the United States Pharmacopoeia.

Surgery was a developed art.

Evidence shows that the ancients

Mayans, like the ancients Egyptians, cut

out sections of bone from the skull of

living persons. The operation, called

trepanning, was performed on

rulers or elite persons. Some patients

lived through the operation, as indicated

by the discoveries of skulls with the

replaced bone healed or partially healed.

We can assume the high status of the

patient because the skulls found also

showed the cosmetic insertion

of jade or green stone into the faces of

the teeth.

Some music we can see in some

Mayan's murals where they are using

instruments like horns, trumpets and

flutes.

Similar Writing and meaning

between Japanese, Old Egyptian and

Nahuatl from Mexico.

In the language of Japanese, Old

Egyptian and Nahuatl from Mexico are

lots of words with the same meaning and

the similar writing.

Example;

Japanese: Tori means bird.

Egyptian: Tot means bird

Nahuatl: Totol means bird

Japanese: Ka means house

Egyptian: Ka means house

Nahuatl: Ka means house

Playing with another languages like

German, English & Maya.

German: Konig means house

English: King means "Who rule the

kingdom, better say king

Maya: Kin means King

The world in Egyptian TUTANKAMON

means the bird that lives in Amon

Tut or Tot=bird

an=   in

ka=   house

Amon=   Amon

The question, why the language of

Japanese, Egyptian & Nahuatl are

similar. Even their faces are similar.

The Rosetta Stone opened a new door to understand the demotic Egyptian symbols and fascinating story about the King of the South and North **PTOLEMY, the ever-living, the beloved of Ptaḥ who gave** Forgiveness' and freedom to slaves .

Ptolmys seal in old Egyptian semiotic symbols , read from right to left

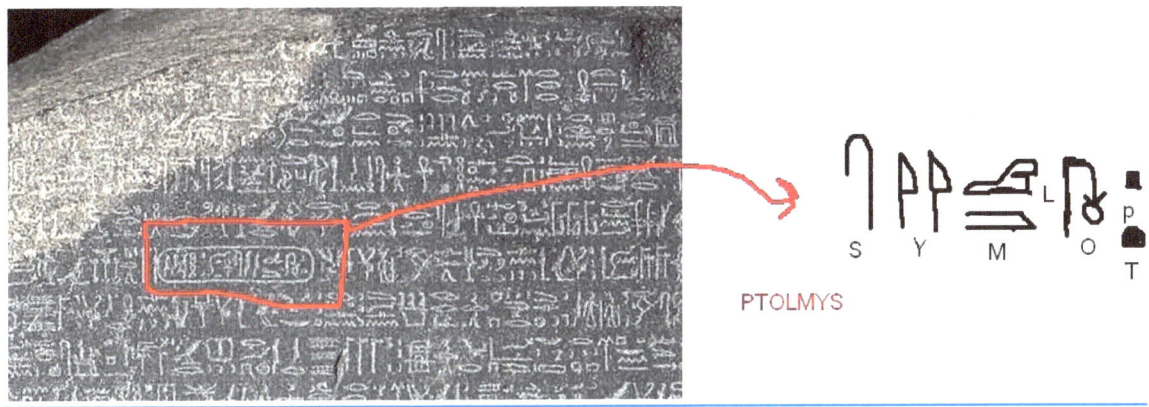

PTOLMYS

PTOLEMY

PTOLOMEO
IN GREEK

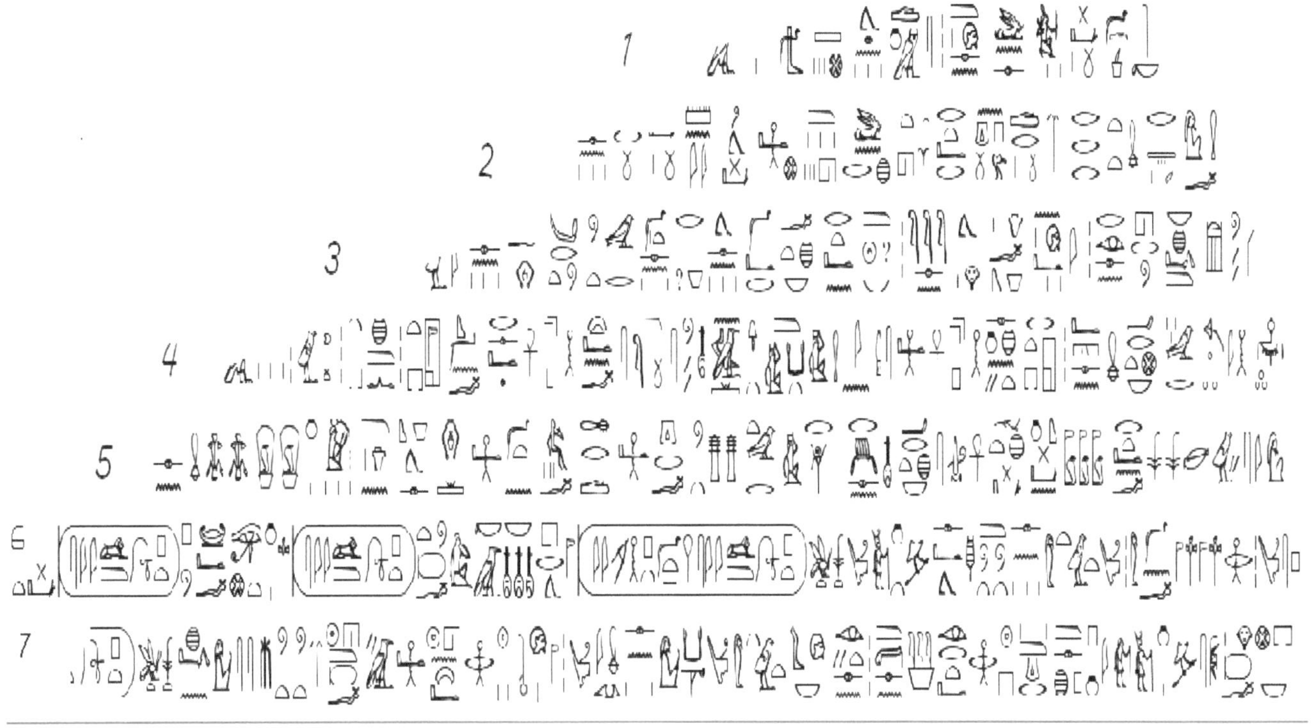

8

9

10

11

12

13

14

The Rosetta Stone was discovered in 1799 near el-Rashid in those days named as Rosetta in Egypt by Napoleon's Army Captain François Xavier Bouchard. he Rosetta stone fragment carried three parallel inscription 2 in Egyptian A) semiotic, hieroglyphic symbols and demotic writing and the third in Greek writing , explain how great was the pharaoh Ptolemy V in 196 b.c.   In 332 B.C Alexander the great conquered Egypt and the language of greek started popular as second language besides Egyptian language. In 205 bc Ptolemy

V reign and the country was in chaos so he made publicity proclaim him self the saivor, of Egypt and one of the publicity tools was the Rosette Stone in 196 bc and be the one to became pharaoh of Egypt.

TRANSLATION OF THE HIEROGLYPHIC TEXT ON THE ROSETTA STONE AND ON THE STELE OF DAMANHÛR. THE DECREE WAS IN THE 9TH YEAR OF THE REIGN OF PTOLEMY V. EPIPHANES.

1. On the twenty-fourth day of the month GORPIAIOS, which corresponded to the twenty-fourth day of the fourth month of the season PERT of the inhabitants of TA-MERT (EGYPT), in the twenty-third year of the reign of HORUS-RA the CHILD, who hath risen as King upon the throne of his father, the lord of the shrines of NEKHEBET and UATCHET, the mighty one of two-fold strength, the stablisher of the Two Lands, the beautifier of

2. Egypt, whose heart is perfect (or benevolent) towards the gods, the HORUS of Gold, who makes perfect the life of the *hamentet* beings, the lord of the thirty-year festivals like PTAḤ, the sovereign prince like RĀ, the King of the South and North, **Neterui-merui-⊡tui-⊡uā-setep-en-Ptaḥ-usr-ka-Rā-ānkh-sekhem-⊡men** , the Son of the Sun **Ptolemy, the ever-living,**

the beloved of Ptaḥ, the god who maketh himself manifest.

3. the son of **PTOLEMY** and **ARSINOË**, the Father-loving gods; when PTOLEMY, the son of PYRRHIDES, was priest of ALEXANDER, and of the Saviour-Gods, and of the Brother-loving Gods, and of the Beneficent Gods,

4. and of the Father-loving Gods, and of the God who maketh himself manifest; when DEMETRIA, the daughter of Telemachus, was bearer of the

5. prize of victory of BERENICE, the Beneficent Goddess; and when ARSINOË, the daughter of CADMUS, was the Basket Bearer of ARSINOË, the Brother-loving Goddess;

6. when IRENE, the daughter of PTOLEMY, was the Priestess of ARSINOË, the Father-loving Goddess; on this day the superintendents of the temples, and the servants of the god, and those who are over the secret things of the god, and the libationers [who] go into the most holy place to array the gods in then apparel,

7. and the scribes of the holy writings, and the sages of the Double House of Life, and the other libationers [who] had come from the sanctuaries of the South and the North to MEMPHIS, on the day of the festival, whereon

S. His Majesty, the King of the South and North **PTOLEMY, the ever-living, the beloved of Ptaḥ**, the god who maketh himself manifest, the lord of beauties, received the sovereignty from his father,

entered into the SEḤETCH-CHAMBER, wherein they were wont to assemble, in MAKHA-TAUI , and behold they declared thus:—

9. "Inasmuch as the King who is beloved by the gods, the King of the South and North **Neterui-merui-⊡tui-⊡ua-en-Ptaḥ-setep-en-usr-ka Rā ānkh-sekhem-⊡men,** the Son of the Sun **Ptolemy, the ever-living, beloved of Ptaḥ**, the Gods who have made themselves manifest, the lord of beauties, hath given things of all kinds in very large quantities unto the lands of Horus and unto all

10. "those who dwell in them, and unto each and every one who holdeth any dignity whatsoever in them, now behold, he is like unto a God, being the son of a God [and] he was given by a Goddess, for he is the counterpart of Horus, the son of Isis [and] the son of Osiris, the avenger of his father Osiris—and behold, His Majesty.

11. "possessed a divine heart which was beneficent towards the gods; and he hath given gold in large quantities, and grain in large quantities to the temples and he hath given very many lavish gifts in order to make Ta-Mert [Egypt] prosperous, and to make stable [her] advancement;

12. "and he hath given unto the soldiers who are in his august service . . . . . . according to their rank

and of the taxes] some of them he hath cut off, and some of them [he hath lightened], thus causing the soldiers and those who live in the country to be prosperous

13. "under his reign [and as regards the sums which were due to the royal house] from the people of Egypt, and likewise those [which were due] from every one who was in his august service, His Majesty remitted them altogether, howsoever great they were;

14. "and he hath forgiven the prisoners who were in prison, and ordered that every one among them should be released from [the punishment] which he had to undergo. And His Majesty made an order saying:—In respect of the things [which are to be given to] the gods, and the money and the

15. "grain which are to be given to the temples each year, and all the things [which are to be given to] the gods from the vineyards and from the corn-lands of the nome, all the things which were then due under the Majesty of his holy father

16. "shall he allowed to remain [in their amounts] to them as they were then; and he hath ordered:—Behold, the treasury (?) shall not he made more full of contributions by the hands of the priests than it was up to the first year of the reign of His Majesty, his holy father; and His Majesty hath remitted

17. "To the priests who minister in the temples in courses the journey which they had been accustomed to make by river in boats to the city of ALEXANDRIA at the beginning of each years and His Majesty commanded:—Behold, those who are boatmen [by trade] shall not be seized [and

made to serve in the Navy]; and in respect of the cloths of byssus [which are] made in the temples for the royal house,

18. "he hath commanded that two-thirds of them shall be returned [to the priests]; similarly, His Majesty hath [re]-established all the things, the performance of which had been set aside, and hath restored them to their former condition, and he hath taken the greatest care to cause everything which ought to be done in the service of the gods to be done in the sane way in which it was done

19. "in former [days]; similarly, he hath donc [all things] in a right and proper manner; and he hath taken care to administer justice *to the people, even like Thoth, the great, great [God]; and he hath, more over, ordered in respect of those of the troops who come back, and the other people also, who during the*

20. "*strife of the revolution which took place had been ill disposed [towards the Government], that when they return to their homes and lands they shall have the power to remain in possession of their property, and he hath taken great care to send* infantry, and cavalry, and ships to repulse those who were coming against

21. "Egypt by land as well as by sea; and he hath in consequence expended a very large amount of money and of grain on them in order to make prosperous the lands of Horus and Egypt.

22. "And His Majesty marched against the *town of Shekam*, which is in front of (?) the town of UISET, *which was in the possession of the enemy, and was provided with catapults, and was made ready for war with weapons of every kind by*

23. "the rebels who were in it—now they had committed great acts of sacrilege in the land of Horus, and had done injury to those who dwelt in Egypt—His Majesty attacked them by making a road [to their town],

24. "and he raised mounds (or walls) against them, and he dug trenches, and whatsoever would lead [him] against them that he made; *and he caused the canals which supplied the town with water to be blocked up, a thing which none of the kings who preceded him had ever been able to do before, and he expended a large amount of money on carrying out the work;*

25. "and His Majesty stationed infantry at the mouths of the canals *in order to watch and to guard them against the extraordinary rise of the waters [of the Nile], which took place in the eighth year [of his reign], in the aforesaid canals which watered the fields, and were unusually deep*

26. "in this spot; and His Majesty captured the town by assault in a very short time, and he cut to pieces the rebels who were therein, and he made an exceedingly great slaughter among them, even like unto that which THOTH ₁ and HORUS, the son of Isis and [the son of Osiris], made among those who rebelled against them

27. "when they rebelled in this very place; and behold, those who had led on the soldiers and were at their head, and who had disturbed the borders [in the time of his father, and who had committed sacrilege in the temples, when His Majesty came to MEMPHIS to avenge his father

28. "and his own sovereignty he punished, according to their deserts, when he came there to celebrate] the festival of the receiving of the sovereignty from his father; and [besides this], he hath set aside [his claim to

29. "the things which were due to His Majesty, and which were [then] in the temples, up to the eighth year [of his reign, which amounted to no small sum of] money and grain; and His Majesty hath also set aside [his claim] to the cloths of byssus which ought to have been given to the royal house and were [then] in the temples,

30. "and also the tax which they (*i.e.* the priests) ought to have contributed for dividing the cloths into pieces, which was due up to this day; and he hath also remitted to the temples the grain which was usually levied as a tax on the corn-lands of the gods, and likewise the measure of wine which was clue as a tax on vineyards [of the gods];

31. "and he hath done great things for APIS, and MNEVIS, and for every shrine which contained a sacred animal, and he expended upon them more than did his ancestors; and his heart hath entered into [the consideration of everything] which was right and proper for them

32. "at every moment; and he hath given everything which was necessary for the embalming of their bodies, lavishly, and in magnificent abundance; and he hath undertaken the cost of their maintenance in their temples, and the cost of their great festivals, and of their burnt offerings, and sacrifices, and libations;

33. "[and he hath respected the privileges of the temples, and of Egypt, and hath maintained them in a suitable

manner according to what is customary and right; and he hath spent] both money and grain to no small amount;

34. "and [hath provided] everything in great abundance for the house wherein dwelleth the LIVING APIS; and His Majesty hath decorated it with perfect and new ornamentations of the most beautiful character always; and he hath made the LIVING APIS to rise [like the sun], and hath founded temples, and shrines, and chapels [in his honour]; [and he hath repaired the shrines, which needed repairs, and in all matters appertaining to the service of the gods

35. "he hath manifested the spirit of a beneficent god; and during his reign, having made careful inquiry, he hath restored the temples which were held in the greatest honour, as was right] and in return for these things the gods and

goddesses have given him victory, and power, and life, and strength, and health, and every beautiful thing of every kind whatsoever, and

36. "in respect of his exalted rank, it shall be established to him and to his children for ever and ever, with happy results (or life)."

And it has entered into the heart(s) of the priests of the temples of the South and of the North, and of each and every temple [that all the honours which

37. are paid] to the King of the South and North **Ptolemy, the ever-living, the beloved of Ptaḥ**, the [God who maketh himself manifest, whose deeds are beautiful, and those which are paid to the Father-loving Gods who begot him, and to the Beneficent Gods who begot those who begot him, and to the Brother-Gods who begot the begetters of his begetters,]

38. and to the Saviour-Gods, shall be [greatly increased]; and a statue of the King of the South and North, **Ptolemy, ever-living, beloved of Ptaḥ**, the God who maketh himself manifest, the Lord of beauties, shall be set up [in every temple, in the most prominent place], and it shall be

39, called by his name "**PTOLEMY, the SAVIOUR of EGYPT**," the interpretation (?) of which is "**PTOLEMY, THE VICTORIOUS ONE**." [And it shall stand side by side with a statue of the Lord of the gods (?), who giveth him the weapon of victory, and it shall be fashioned after the manner of the Egyptians, and a statue of this kind shall be set up in]

40. all the temples which are called by his name. And adoration shall he paid unto these statues three times each day, and every rite and ceremony which it is proper to perform before them shall be performed, and whatsoever is prescribed, and is fitting for their DOUBLES, shall be performed, even as it is performed for the gods of the Nomes during the festivals and on every sacred day (?), on the day of [his] coronation, and on his name-day. And there shall likewise [be set up] a

41. magnificent (?) statue of the King of the South and North **Ptolemy, ever-living beloved of Ptaḥ**, the God who maketh himself manifest, whose deeds are beautiful, the son of **Ptolemy**, and **Arsinoë**, the Father-loving gods, and with the statue there shall be a magnificent shrine [made] of the finest copper and inlaid with real stones of every kind,

42. in every temple which is called by his name; and this statue shall rest in the most holy place [in the temples] side by side with the shrines of the gods of the Nomes. And on the days of the great festivals, when the god [of the temple] cometh forth from his holy habitation, according to his day, the holy shrine of the God who maketh himself manifest, the lord of beauties, shall likewise be made to rise [like the Sun]

43. with them. And in order to make this new shrine to be easily distinguishable [both at the present day, and in future times, they shall set] upon this shrine [ten royal double crowns, made of gold and upon [each of the double crowns there shall be placed the [serpent which it is right and proper to make for the [double crown of gold], instead of the two Uraei

44. which are [placed] upon the tops of the shrines, and the SEKHENT CROWN shall be in the middle of them, because it was in the SEKHENT CROWN in which His Majesty shone in the house of the KA of PTAḤ (*i.e.*, Memphis)

45. at the time when the king entered into the temple, and performed the ceremonies which it was meet and right for him to perform on receiving the exalted rank [of King]. And on the upper surface of the square pedestal which is round these crowns, and in the middle part thereof [which is immediately] beneath] the double Crown [*they shall engrave a papyrus plant and a plant of the*

south; and they shall set them in such a way that a vulture, *upon neb,* *, beneath which a plant of the south shall be found, shall be affixed to the*

right-hand upper corner of the golden shrine, and a serpent, *, under*

which is *, placed upon*] a papyrus plant, [shall be affixed] to the left hand side [at the upper corner]; and

46. the interpretation [of these signs is]:—"Lord of the shrine of NEKHEBET, and Lord of the shrine of UATCHET, who illumineth the land of the White Crown, and the land of the Red Crown." And inasmuch as the last day of the fourth month of the season SHEMU (*i.e.*, MESORE), which is the birthday of the beautiful ever-living god, is already established as a feast day, and it hath been observed as a day of festival in the lands of HORUS (*i.e.*, the temple lands)

from the olden time; and moreover, the seventeenth day of the second month of the season SHAT (i.e., PAOPI),

47. whereon [His Majesty] performed the ceremonies of royal accession, when he received the sovereignty from his father, [is also observed as a day of festival], and behold [these days] have been the source of all [good] things wherein all men have participated; these days, that is to say, the seventeenth and the last day of each month, shall be kept as festivals in the temples

48. of Egypt, in each and every one of them; and on these days burnt offerings shall be offered up, and meat offerings, and everything which it is right and customary to perform at the celebration of festivals shall be performed on these days every month, and on these festivals every man shall do (i.e., offer up) what he is accustomed to do on [other] fes-

49. tivals in the temples. [And the priests also decreed] *that the things which [are brought to the temples]* as

offerings shall be given unto the persons who [minister in the temples; and festivals and processions shall be established in the temples, and in all Egypt, in honour of] the King of the South and North, **Ptolemy, ever-living, beloved of Ptaḥ,** the god who maketh himself manifest, whose deeds are beautiful, each year,

50. beginning with the first day of the first month of the season Shat (i.e., Thoth) up to the fifth day thereof [and on these days the people shall wear] garlands on their heads, and they shall make festal the altars, and shall offer up meat and drink offerings, and shall perform everything which it is right and proper to perform. And the priests of all the temples which are called after his name

51. shall have, in addition to all the other priestly titles which they may possess, the title of "Servant of the god who maketh himself manifest, whose deeds are beautiful"; [*and this title shall be endorsed on all deeds and documents which are laid up in the temples*]; and they shall cause to be engraved on the rings which they wear on their hands, the title of "Libationer of the god who maketh himself manifest, whose deeds are beautiful."

52. And behold, it shall he in the hands of those who live in the country, and those who desire [it], to establish a copy of the shrine of the god who maketh

himself manifest, whose deeds are beautiful, and set it up in their houses, and they shall be at liberty to keep festivals and make rejoicings [before it] each month

53. and each year; and in order to make those who are in Egypt to know [*why it is that the Egyptians pay honour—as it is most right and proper to do—to the*

god who maketh himself beautiful, whose deeds are beautiful, the priests have decreed] that this DECREE shall [*be inscribed*] upon a stele of hard stone in the writing of the words of the gods, and the writing of the books, and in the writing of HAUI-NEBUI (i.e., Greeks), and it shall be set up in the sanctuaries in the temples which [are called] by his name, of the first, second, and third [class], near the statue of the HORUS, the King of the South and North **Ptolemy, ever-living, beloved of Ptaḥ**, the god who maketh himself manifest, whose deeds are beautiful.

Hieroglyphic Alphabet

a		h		o		v	
b		i		p		w	
c		j		q		x	
d		k		r		y	
e		l		s		z	
f		m		t		boy	
g		n		u		girl	

Probably these civilizations were

together as one continent before the

big flood arrive. In ancients stories they

mention the name of Noah in the Bible

(o' biblws, that means The Book) , a

humble man chosen by the God of

Abraham.

Noah build an arc by the instructions

of God for the big flood that will be

vanished the world.

If we check the name of Noah and we

move the letters of Noah a little we have

Nahoa Well, I'm playing to much with

the names, but think about it.

There is a legend in the Nahoas (Old

Mexican culture) that says in the book of

Bernal Diaz Del Castillo;

····.And the Nahoas arrive to the Indian continent through Panico-Panitla Veracruz.. In his ancients stories they survived a big flood that destroyed his old world ..and they called them Atlantes….

Bernal Diaz Del Castillo was a soldier

and writer of Hernan Cortez in 1517..

Thanks for Bernal Diaz Del Castillo we

have more legends from the Nahoas…

There is another amazing legend

describe by Bernal Diaz Del Castillo.

Papantzin sister of Moctezuma was sick and had a nightmare where white people from the sea will arrived and conquered the Nahoas in Mexico.

This dream that had Papantzin was

before the Spanish with Hernan Cortez

came to Mexico.

Well, before the big flood, the continents

were together so there were few

languages.

Probably there was an old language

were Japanese, Nahuatl, and Egyptian

born.

The continents of the world are like a

puzzle. America can link with the

Asiatic continent and the African

continent as well.

While we study history, we understand

more about us. The first time I read

Marco Polo from 14th century, I learned

amazing things, like the Piñata (Clay

container dressed with paper). Marco

Polo was a great voyager, but

accidentally he started to travel. Once

upon a time Marco Polo was in a night

bar and someone was killed and

everybody pointed Marco Polo. Marco

Polo was sentenced to death by the Holy

Inquisition, but his uncle Mafio was a

member of the Signori Dela Notti so

Marco Polo was banished.

Marco Polo with his father and his uncle

started to travel to Orient. When they

arrived to Orient with the Chinese

emperor Kan Kubilai in the center of the

place there were people hitting a clay

container with a stick. When the Chinese

people break the clay, lots of

fruits and seed start falling to the floor.

This ritual means purification of the

soul.

Marco Polo brought this ritual to Italy

and there some one called it Piñata, then

past to Spain and in Spain someone

dressed the Piñata. When the Spanish

arrived in 1517 to Acolman, Mexico

they saw the Natives Nahoas hitting a

clay container with lot of fruit

and seeds with the same ritual like in

Chine from 14th century. The question,

who brought the tradition of the piñata to

America? Maybe the Vikings or the old

Nahuatl, better say

Nahoa or Noah.

The Noah's arc, a legend, a story or fairy

tail. Well, we can hear this story in the

Old Testament in the Bible, that God

told Noah to build an arc and pick up

selected animals.

God told Noah that will make a big flood

all over the planet so only he and his

family will survive. For 40 days and

40 nights was rainy so the big flood

came.

There is another legend from the

Sumerians and was the epic of

Gilgamesh. Gilgamesh was a giant that

was pushing an arc from a big flood.

Gilgamesh save all the crew from

the big boat or better say arc.

The scientist now are agree that

in the world was a big flood like

10 thousand years ago. Well, the cause

of this big flood, the continents were

separated.

If we see a map, there is a lot of

coincidence between the pyramids of

Egypt and Mexico's pyramids, they are

in the same 30º degrees longitude in a

mapa mundi..

Sometimes we think that a new language

to learn is going to be difficult to speak.

Sometimes we think that the languages

were created by wise people. But, not,

the ignorance are the ones to change

them. In America all the languages that

we speak are depuration from the

original languages from Europe. The

French from France is the original
French to speak. The French from
Canada is not the original is kind of the
depuration . Like the English from
England is beautiful and unique, but the
English in America is different.
For accident the Mexicans speak
Spanish, well to understand this we have
to back in time to the 16 century. In
those days the Pope from the Catholic
Church has to divide the world because
the king of Spain & Portugal were in
war. So the Pope ask to the king of
Spain which part of the world would like
to have, and Spain choose from center of
America to the North America, so
Portugal choose the South. Well, the
Pope let choose Spain first because
Spain discovered America first.
The big surprise that the king of Spain
received was that he didn't know that
South America was so big. So that's why

in Mexico the language is Spanish and in

Portugal is Portuguese.

If we check the French language we

found a beautiful and romantic language.

There are words like to say "Father in

law", we say in French " Le beau pere" (

The good father).

"Sister in law" we say "La Belle soeur "

(The beautiful sister). Well the language

of French is so romantic. And the

pronunciation is so unique.

But, the true story is that the new French

that is spoken in France is not the real

French from the 15th century.

In The 15th century, the language of

French was spoken like latin, but the

ignorance changed it so romantic and

beautiful.The ignorance eliminated

letters like "S" in many words, example

of this we find in theWord Isle, so they

eliminated the "S" and they replaced it

with the symbol of circumflex accent " ^

" so the word Isle they wrote it as île.

So today the word island in

French we write as île. They also

eliminated vowels adding apostrophe

like the sentence

La île, they changed as L' île etc…

The English that we speak today is the

mix of Latin & Old German, the story

mentioned that the Old Germans or

Normans arrived to the British Islands in

the 4 century and the Old German mixed

with Latin and English was born.

Then another 9 languages like German,

French, Hollandaise, Flamenco, Danish,

Norwedish, Polish, Spanish & English.

So, what kind of language we are talking

today, what kind of new word we

created by accident or lack of

knowledge. At the end of this quietness

of sounds and new words,

there is the night that come and a new

dialect or language born.

"Final Verdict, Abraham Lincoln Election"

Abraham Lincoln Election

1. In February 1862, while Mary Lincoln hosted a dinner reception downstairs, her 12 years old son, Willie, lay dying upstairs.

2. In the reception we can see The Treasure Secretary Salmon Chase shaking hand with Lincoln and his wife Mary Lincoln.

Lincoln:

Is a pleasure that you came to our reception, in addition congratulation for your triumph of became treasure secretary.

Salmon Chase:

I present to you my daughter Kate and her husband William Sprague.

Lincoln:

Is an honor meet such beautiful lady and her lucky husband.

(We can see William H Seward approach to Lincoln)

William H. Seward:

Hi Mr President what a nice reception !

Lincoln:

 In deed.

3. We can see in the reception night party also John Hay and John Nicolay talking

John Hay:

A lot of beautiful ladies tonight.

John Nicolay:

You always with a good sense of humor.

Continuing ...in the reception night party

John Hay:

 Who is that beautiful lady over there.

John Nicolay:

Well she is Kate, is married with Mr. William Sprague. There are rumors that Mr. William Sprague financed her father Salmon Chase Campaign in 1864 so Salmon Chase became treasure secretary.

John Hay:

That's interesting.

John Hay:

What are you drinking?

John Nicolay:

Red Wine, hey by the way your girlfriend is coming over there

(We can see a beautiful lady in her 20's approaching to John Hay her name is

Clara Louisa Stone:

Where you've been?

John Hay:

I've been here all the time, waiting for you darling, i present you John Nicolay

John Nicolay:

you always kidding

Nice to meet you young lady.

(We can see John Hay and his girlfriend Diana Smith talking)

Clara Louisa Stone:

I need to speak in private.

John Hay:

You can talk here

Clara Louisa Stone:

Do you love me?

John Hay:

Of course i love you, but I've been pretty busy working with the president Lincoln

(John Hay start kissing each other John Hay and Clara Louisa Stone)

Clara Louisa Stone:

I thought you don't care about me

(John Hay grab her again and start kissing her)

John Hay:

I'm so sorry !

What would you like to drink?

Clara Louisa Stone:

White wine.

John Hay:

You look beautiful tonight

Clara Louisa Stone:

come on follow me

(Clara Louisa Stone grab John Hay to a part of the reception party quite alone and kiss him a little more)

then the newspaper from N.Y Samuel Week takes a pictures of John Hay and Clara Louisa Stone kissing each other.

Then they come back to the reception party

let me introduce you with Kate

(

We can see John Hay introducing his fiancée Clara Louisa Stone to Kate Sprague)

Kate she is my fiancée Clara

Kate Sprague:

Nice to meet you

Where are you from?

Clara Louisa Stone:

I from Chicago.

Kate:

is really cold there in winter time isn't?

Clara Louisa Stone:

In deed.

Kate :

How you meet John Hay?

Clara Louisa Stone:

In a party, all drunk

Kate:

That's usual.

Clara Louisa Stone:

Wow you meet him

Kate:

Through my father.

(Then General Edward Bates approach to the Lincoln's secretaries John Hay with his fiance Clara Louisa Stone next to Kate and John Nicolay.)

A. General Edward Bates:

It's a pleasure meet the young secretaries of Lincoln's cabinet. What are you drinking Nicolay?

John Hay:

Well he is drinking red wine.

(John Hay and John Nicolay looks each other in the funny way)

(we can see Samuel Weed journalist from New York Times approaching to interview John Hay)

Samuel Weed New York times journalist:

 Excuse me Mr. Secretary Hay, do you understand the situation of the soldiers of this nation in the way that sometimes they don't have boots or jacket to go to war..

John Hay:

 You are right... Is so sad that there was not to much budget for assistant the soldiers.

(In that moment Salmon Chase with a glass of wine in his right hand approach to the conversation)

Salmon Chase:

 Pardon me for the interruption

Samuel Weed New York Times writer.:

 Mr Chase tell me about Mr. Stanton when was tormented by the long lines of ambulances that rolle into the city each morning carrying the injured and the dead from the Peninsula Campaign.

Salmon Chase:

Well , indeed in fact all his life, Stanton had been unnerved in the

presence of death.

Sometimes he took it upon himself to deliver the news to stricken families.

Samuel Weed, N.Y times writer:

 Tell me about Ellet Cabell, whose father , colonel Charles Ellet, was fatally wounded in Memphis, long recalled the moment when Stanton Appeared at her family's home in Georgetown to tell of Ellet's heroism during the battle

John Hay:

Pardon interrupt you, I have heard that this powerful war minister was harsh and feeling but I can never forget the tenderness of his manner as he delivered the news with tears to his eyes.

Stanton's own family was touched in death as well. While his cabinet reeled in the aftermath of the Peninsula defeat, Lincoln was face with the grim knowledge that the ultimate authority had been his alone. None the less , as Whitman had observed following the debacle at Bull Run Lincoln refused to surrender to the gloom of defeat...

In addition while the battle was still ongoing, Lincoln had found time to write a letter to a young cadet at West Point, the son of Mary's cousin Ann Todd Campbell. The boy was miserable at the academy and his mother was worried.

"Allow me to assure you it is a perfect certain that you will very soon, feel better quite happy If you only stick to the resolution you taken to procure a military education.

Samuel Weed New York Times writer:

Tell me about the Ariel early of July 8, 1862 accompanied by assistant secretary of war Peter Watson and congressman Frank Blair.

John Hay:

Accompanied by assistance Secretary of war Peter Watson and congressman Frank Blair , he left Washington aboard the Ariel early on that morning of July 8 1862…

⋯ July 8, 1862 beginning the 12 hr journey to Mc Clellan's new headquarters at flarrison's landing on the James River.

"The day had been intensely hot" an army correspondent noted, the temperature climbing to over 1oo degrees. Even soldiers who lay in the shade of the trees found small respite from the almost over powering heat. By 6 p.m , however, when General Mc Clellan and his staff met the president at Harrison's landing, the setting son had yielded to a pleasant, evening.

News of the president's arrived spread quickly through the camp.

Soldiers in the vicinity let out great cheers whenever they glimpsed him "Sitting and smiling serenely on the after deck of the vessels" Lincoln's calm visage , however marked his deep anxiety about Mc Clellan and the progress of the war.

Imagine Mc Clellan writing a strong frank letter to Lincoln before Lincoln arrived.

Equally troubled the defeated Mc Clellan had spent the hrs. before Lincoln's arrived drafting what he termed . Mc Clellan had the letter to Lincoln, who read it as the two sat together on the deck.

He was relieved to find the army in such high spirits after the bloody week long battle, which had decimated their ranks, leaving 1,734 dead and 8,066.

(We can see when antislavery leader Frederick Douglas say hello to Samuel Weed N.Y Times writer)

Samuel Weed N.Y. times writer.:

What you can tell us about black people join the army.

Frederick Douglas:

More than 200,000 black men served to the army in this country.

(we can see a the distance Senator Chandler of Michigan called McClellan a liar and coward)

Senator Chandler of Michigan:

 You are liar and coward Mr. McClellan.

(John Hay separate them immediately. And David Davis approach and speak with John Hay)

3.A continuing in reception

David Davis:

What such violence !

John Hay:

In deed sir, is a pleasure see you in our reception, the president Lincoln he mentioned a lot about your friendship with him..

David Davis:

Well since I remember I join his steadfast companion when I from my office at the circuit court judge for the 8th district to secure Lincoln's nominations. There was also Norman Judd, an attorney for the railroads and chairman of the Illinois republican state central committee; Stephan Logan, Lincoln's law partner for 3 years in the early 40's . Lincoln and me we had good and bad times that we struggle as a lawyers shared rooms and sometimes beds in dusty village inns and taverns spending long evening gathered together around a blazing fire.

John Hay:

Now I'm understand why all this is called "The Circuit" studying thousands of small cases in order to make a living. The arrival of the traveling bar brought life and vitality to the county seats, fellow rider Henry Whithey recalled, When the court session finished.

David Davis:

In Chicago this was clearly understood by Lincoln's team in Chicago named "The Circuit"

John Hay:

Tell me about Nathan Knapp

David Davis:

Well Nathan Knapp told Lincoln when he first arrived in Chicago.

"Keep a good nerve" Knapp advised be not surprised at any result but I tell your chances are not the worst.

John Hay:

So in that way Mr. Seward was not nominated on the first ballot., it was his judgment that Mr. Salmon Chase of Ohio or Mr. Bates of Missouri would be the nominee

David Davis:

Cutting reasons why each of those two candidates would have difficulty securing the nomination.

I can tell you that friends are forever sharing good and bad times, sometime sharing beds like Lincoln and Joshua Speed companion when they had similar dispositions both. No longer a boy but not yet an establish adult , Lincoln ended years of emotional deprivation and intellectual solitude by building his first and deepest friendship with Speed.

John Hay:

So Mr. Joshua Speed was most intimate friend of Abraham Lincoln.

David Davis:

In addition I can tell you that Lincoln before his marriage enjoyed close relations with young women and almost certainly found outlets for his sexual urges, among the prostitutes who were readily available on the front.

John Hay:

Mr. President Lincoln always devoted to his law carrier

Davis David:

Well Mr. President Lincoln always busy, even when his wife Mary reporting nursing troubles with Willie, and conversation in which Lincoln had confided that both he and Mary were hoping for a girl before Tad was born. Nothing in recently letters were marital discord in the Lincoln home.

John Hay:

In deed sir Davis.

Davis David: I heard that you Mr. Hay were a journalist from Missouri Democrat Newspaper.

John Hay: Yes sir I was lucky to be accepted, in fact I graduate at Brown University and they helped me with good recommendation..

Davis David: Besides all this, what you like to do?

John Hay: I love writing poems.

David Davis: How you became one of the secretaries from the President Abraham Lincoln.

John Hay: Well, John Nicolay recommended me with the President Abraham Lincoln, well Mr. Davis I have to keep going, was delightful talking to you.

Fade in/Fade out

4-   Is fall in the year of 1862, after few years after when Abraham Lincoln won election as president of United States of America. We can see John Hay his secretary walking with him along down town in Washington D.C.

John Hay:

Mr. President would you like to share with me when you were young.

Abraham Lincoln:

Well , it's long time ago but I can tell you that when I was your age ,Buying books was difficult for me so I have to go to the school and

                Study there. I remember one time I used to read a lot the Bible and

                Shakespeare. Reading page after page while my horse rested at the

                Craford, a well- to do farmer who lived sixteen miles away. At night

                , where by the light of a tallow candle, or if tallow was scarie, by a

                grease lamp made from hickory bark gathered in

the woods, I read as   long as I could stay awake, placing the book on a makeshift shelfbetween the cabin logs so I could I retrieve at day break.

5-We can see Abraham Lincoln walking fast through the rain with his book wet

(FLASH BACK IN THIS ESCENE)

During a severe rainstorm on night the book was badly soiled and the covers warped.

Lincoln:

I went to Crawford's house, explained what had happened, and offered to work off the value of the book. Crawford calculated the value of two full days work pulling corn, which I considered an unfair reimbursement Nevertheless, he straightway set to work and kept blade left on a stalk. Then, having paid his debt, if rote poems and songs lampooning "Josiah blowing his bugle "Crawford's large nose. Thus Crawford, in return for laning me a book and then exorbitantly penalizing me, a permanent,if unflattering

Fade In/ Fade Out

6.We can see again John Hay walking with Abraham Lincoln in downtown Washington D.C.

John Hay:

What about ladies friends, did you have some?

Abraham Lincoln:

Well, in 1837 at Salem Illinois after studying law I used to go out   with Ann Rutledge, I admit that was my first love. I used to see he in his father's tavern

7. Flash back, we can see Abraham Lincoln in his 20's with his girlfriend Ann Rutled inside a tavern at night

Ann Rutled:

Is a beautiful night

A. Lincoln:

In deed it is

Ann R.:

Tell me more about your family, you are so shy to tell me about it

A. Lincoln:

Well I love talk more with you, than talk about myself.

(we can see through a candle from the tavern when Abraham kiss with a lot of love

Ann Rutled)

FADE IN/OUT

We go back escene with John Hay and Abraham Lincoln walking in Washington D.c downtown.

John Hay:

And then what happened?

Abraham Lincoln:

She died so I divulged my feelings to my old friend Isaac Cogda

··· in deed she was a beautiful lady-would have made a good loving

Wife…. I did honestly I truly love the girl and think often…often of

Her now. At New Salem, Ann was a few years younger than me, she

Had blue eyes large and expressive. Her intellect was said to be

"quick-sharp deep and philosophic as well as brilliant.

John Hay:

How she died?

Abraham Lincoln:

Ann was only 22 in the summer of 1835 while New Salem

Sweltered through on of the hottest summers in the history of the

State, a deadly fever possibly typhoid spread through the town. Ann

As well as some of my friends , perished in the epidemics.

8. We can see people in the funeral ceremony cementery memorial of Ann Rutled who died of typhoid. There is Abraham Lincoln and his friends really sad. Is a ranny day so we can feel the sadness and wind blowing through the people and trees.

And we can see an inscription that says….

    "I am Anne Rutledge who sleep

Beneath these weeds,

Beloved in life of Abraham Lincoln,

Wedded to him, not through Union,

But through separation.

Bloom forever, o republic,

From the dust of my bosom.

"Edgar Lee Masters, Spoon River Anthology…."

FADE IN/OUT

9. We can see again John Hay and Abraham Lincoln walking and a photographer from aa local newspaper taking pictures of them.

Abraham L.

Don't be shy let those people take you photos

John Hay:

Okay.

(We can see the face of John Hay with such shyness, with no smile at all.

10.Next day we can see the Abraham Lincoln's secretaries John Nicolay and John Hay walking a long the streets of Down town Washington D.c and then they go inside a tavern and get some drinks and keep talking…

John Nicolay:

Well the struggle to open the door for black recruits had finally ended

when president Lincoln's emancipation proclamation flatly declared

that blacks would be received in to the armed service o the United States.

John Hay:

In addition Stanton authorized Massachusetts governor John Andrew to raise two regiments of black troops.

John Nicolay:

In deed sir Hay, since Massachusetts had only a small black population, Andrew called on Major George L. Stearns to head a recruitment effort that would reach in to New York and other Nothern States. Stearns approached Frederick Douglas for help.

John Nicolay:

Did you hear about the headlines of the press-magazine?

John Hay:

Which magazine?

John Nicolay:

Douglas was overjoyed . He had long believed that the war would not be won so long as the North refused. To employ the black man's arming suppressing the rebels. " He wrote stirring appeals in his Monthly magazine and traveled through out the North, speaking at large meetings in Albany, Syracuse, Buffalo, Philadelphia, and many other cities, offering a dozen answers to the question. Why should a clolored man enlist?

Nothing, he assured them could more clearly legitimize their call for equal citizenship.

John Hay:

In deed , you will stand more erect, walk more and of America may claim America as his country- and have that claim respected.

John Nicolay:

The black soldiers who initially answered Douglass call became part of the famed 54 th Massachusetts regiment. Captained by Robert Gould Shaw, the son of wealthy Boston abolitionist, this first black regiment from the north. Included two of Frederick Douglass's own sons, Charles and Lewis. On May 28, thousands of Bostonians poured in to the streets cheering the men as they marched past the state house of common.

11.Ext/Mexico city Zocalo/day

Exterior Mexico city we can see the president of Mexico Benito Juarez year dec 1863 and John Hay talking.

John Hay: President Benito Juarez from Mexico, is a pleasure to met you, it was so sad that you couldn't show up to the reception from Abraham Lincoln in the white house.

Benito Juarez: I'm really sorry, but i was pretty busy with some issues in Mexico.

John Hay: What i learned mr. president Abraham Lincoln likes writing you letters because you have the same thoughts about freedom to the slaves people and freedom of speech and religion..

Benito Juarez: Well what i like most from president Linconln is that he carry a Bible with him all the time.. And we like the idea of eliminate the abuse of human rights such slavery people..

John Hay: Well, in United States of North America for first time slaves are free and mr President Abraham Lincoln started with that idea risking his life.

In addition before i became one of the secretaries of mr. Abraham Lincoln i was a newspaper journalist and i used to write about his ideas of freedom of speech and rights to people specially african american.

By the way you have a beautiful country.

Benito Juarez: Thanks..Mexico is a beautiful place with pretty culture. Recently in 1824 we independence from Spain. Its was hard, but at the end we kick out the Spanish from Mexico..

Kate Sprague................play by.................

3. We can see in the reception night party also John Hay and John Nicolay talking

John Hay:

A lot of beautiful ladies tonight.

John Nicolay:

You always with a good sense of humor.

Continuing ...in the reception night party

John Hay:

 Who is that beautiful lady over there.

John Nicolay:

Well she is Kate, is married with Mr. William Sprague. There are rumors that Mr. William Sprague financed her father Salmon Chase Campaign in 1864 so Salmon Chase became treasure secretary.

John Hay:

That's interesting.

John Hay:

What are you drinking?

John Nicolay:

Red Wine, hey by the way your girlfriend is coming over there

(We can see a beautiful lady in her 20's approaching to John Hay her name is

Clara Louisa Stone:

Where you've been?

John Hay:

I've been here all the time, waiting for you darling, i present you John Nicolay

John Nicolay:

you always kidding

Nice to meet you young lady.

(We can see John Hay and his girlfriend Clara Louisa Stone talking)

Clara Louisa Stone:

I need to speak in private.

John Hay:

You can talk here

Clara Louisa Stone:

Do you love me?

John Hay:

Of course i love you, but i've been pretty busy working with the president Lincoln

(John Hay start kissing each other John Hay and Clara Louisa Stone)

Clara Louisa Stone:

I thought you don't care about me

(JOhn Hay grab her again and start kissing her)

John Hay:

I'm so sorry !

What would you like to drink?

Clara Louisa Stone:

White wine.

John Hay:

You look beautiful tonight

Clara Louisa Stone:

come on follow me

(Clara Louisa Stone grab John Hay to a part of the reception party quite alone and kiss him a little more)

then the newspaper from N.Y Samuel Week takes a pictures of John Hay and Clara Louisa Stone kissing each other.

Then they come back to the reception party

let me introduce you with Kate

(

We can see John Hay introducing his fiancée Clara Louisa Stone to Kate Sprague)

Kate she is my fiancée Clara

Kate Sprague:

Nice to meet you

Where are you from?

Clara Louisa Stone:

I from Chicago.

Kate:

is really cold there in winter time isn't?

Clara Louisa Stone:

In deed.

Kate :

How you meet John Hay?

Clara Louisa Stone:

In a party, all drunk

Kate:

That's usual.

Clara Louisa Stone:

how you meet him?

Kate:

Through my father.

 Fade In/ Fade Out

6.We can see again John Hay walking with Abraham Lincoln in downtown Washington D.C.

John Hay:

What about ladies friends, did you have some?

Abraham Lincoln:

Well, in 1837 at Salem Illinois after studying law I used to go out   with Ann Rutledge, I admit that was my first love. I used to see he in his father's tavern

7. Flash back, we can see Abraham Lincoln in his 20's with his girlfriend Ann Rutled inside a tavern at night

Ann Rutled:

Is a beautiful night

A. Lincoln:

In deed it is

Ann R.:

Tell me more about your family, you are so shy to tell me about it

A. Lincoln:

Well I love talk more with you, than talk about myself.

(we can see through a candle from the tavern when Abraham kiss with a lot of love

Ann Rutled)

FADE IN/OUT

We go back escene with John Hay and Abraham Lincoln walking in Washington D.c downtown.

John Hay:

And then what happened?

Abraham Lincoln:

She died so I divulged my feelings to my old friend Isaac Cogda

··· in deed she was a beautiful lady-would have made a good loving

Wife…. I did honestly I truly love the girl and think often…often of

Her now. At New Salem, Ann was a few years younger than me, she

Had blue eyes large and expressive. Her intellect was said to be

"quick-sharp deep and philosophic as well as brilliant.

John Hay:

How she died?

Abraham Lincoln:

Ann was only 22 in the summer of 1835 while New Salem

Sweltered through on of the hottest summers in the history of the

State, a deadly fever possibly typhoid spread through the town. Ann

As well as some of my friends , perished in the epidemics.

8. We can see people in the funeral ceremony cementery memorial of Ann Rutled who died of typhoid. There is Abraham Lincoln and his friends really sad. Is a ranny day so we can feel the sadness and wind blowing through the people and trees.

And we can see an inscription that says….

"I am Anne Rutledge who sleep

Beneath these weeds,

Beloved in life of Abraham Lincoln,

Wedded to him, not through Union,

But through separation.

Bloom forever, o republic,

From the dust of my bosom.

"Edgar Lee Masters, Spoon River Anthology…."

(In a moment David Davis start talking with John Hay)

3.A continuing in reception

David Davis:

What such violence !

John Hay:

 In deed sir, is a pleasure see you in our reception, the president Lincoln he mentioned a lot about your friendship with him..

David Davis:

Well since I remember I join his steadfast companion when I from my office at the circuit court judge for the 8th district to secure Lincoln's nominations. There was also Norman Judd, an attorney for the railroads and chairman of the Illinois republican state central committee; Stephan Logan, Lincoln's law partner for 3 years in the early 40's . Lincoln and me we had good and bad times that we struggle as a lawyers shared rooms and sometimes beds in dusty village inns and taverns spending long evening gathered together around a blazing fire.

John Hay:

Now I'm understand why all this is called "The Circuit" studying thousands of small cases in order to make a living. The arrival of the traveling bar brought life and vitality to the county seats, fellow rider Henry Whithey recalled, When the court session finished.

David Davis:

In Chicago this was clearly understood by Lincoln's team in Chicago named "The Circuit"

John Hay:

Tell me about Nathan Knapp

David Davis:

Well Nathan Knapp told Lincoln when he first arrived in Chicago.

"Keep a good nerve" Knapp advised be not surprised at any result but I tell your chances are not the worst.

John Hay:

So in that way Mr. Seward was not nominated on the first ballot., it was his judgment that Mr. Salmon Chase of Ohio or Mr. Bates of Missouri would be the nominee

David Davis:

Cutting reasons why each of those two candidates would have difficulty securing the nomination.

I can tell you that friends are forever sharing good and bad times, sometime sharing beds like Lincoln and Joshua Speed companion when they had similar dispositions both. No longer a boy but not yet an establish adult , Lincoln ended years of emotional deprivation and intellectual solitude by building his first and deepest friendship with Speed.

John Hay:

So Mr. Joshua Speed was most intimate friend of Abraham Lincoln.

David Davis:

In addition I can tell you that Lincoln before his marriage enjoyed close relations with young women and almost certainly found outlets for his sexual urges, among the prostitutes who were readily available on the front.

John Hay:

Mr. President Lincoln always devoted to his law carrier

Davis David:

Well Mr. President Lincoln always busy, even when his wife Mary reporting nursing troubles with Willie, and conversation in which Lincoln had confided that both he and Mary were hoping for a girl before Tad was born. Nothing in recently letters were marital discord in the Lincoln home.

John Hay:

In deed sir Davis.

Davis David: I heard that you Mr. Hay were a journalist from Missouri Democrat Newspaper.

John Hay: Yes sir I was lucky to be accepted, in fact I graduate at Brown University and they helped me with good recommendation..

Davis David: Besides all this, what you like to do?

John Hay: I love writing poems.

David Davis: How you became one of the secretaries from the President Abraham Lincoln.

John Hay: Well, John Nicolay recommended me with the President Abraham Lincoln, well Mr. Davis I have to keep going, was delightful talking to you.

Fade in/Fade out

FADE IN/OUT

11.Ext/Mexico city Zocalo/day

Exterior Mexico city we can see the president of Mexico Benito Juarez year dec 1863 and John Hay talking.

John Hay: President Benito Juarez from Mexico, is a pleasure to met you, it was so sad that you couldn't show up to the reception from Abraham Lincoln in the white house.

Benito Juarez: I'm really sorry, but i was pretty busy with some issues in Mexico.

John Hay: What i learned mr. president Abraham Lincoln likes writing you letters because you have the same thoughts about freedom to the slaves people and freedom of speech and religion..

Benito Juarez: Well what i like most from president Linconln is that he carry a Bible with him all the time.. And we like the idea of eliminate the abuse of human rights such slavery people..

John Hay: Well, in United States of North America for first time slaves are free and mr President Abraham Lincoln started with that idea risking his life.

In addition before i became one of the secretaries of mr. Abraham Lincoln i was a newspaper journalist and i used to write about his ideas of freedom of speech and rights to people specially african american.

By the way you have a beautiful country.

Benito Juarez: Thanks..Mexico is a beautiful place with pretty culture. Recently in 1824 we independence from Spain. Its was hard, but at the end we kick out the Spanish from Mexico..

LETRAS DE DIOSES MEXICANOS

Words from Mexican Gods

Written by Erik De La Torre Stahl copyrights 1990

In this novel written by Erik De La Stahl invite you to know some of the amazing writers and poets from Mexico in the 1870's. This is the story of an amazing poet named Manuel Acuna in love with Rosario De La Pena

1. Int /Casa de la Familia de Rosario De La Pena hoy El Palacio de Bellas Artes./atardecer

Is the year 1873.

Is the year 1873 and is almost sunrise in the house Rosario De La Pena today Theatre Palacio de Bellas Artes.

)

There is a great party in the ancestors family from Rosario De La Pena she is 26 years old her father Juan

We can see as well the teacher Gabino Barreda, El Nigromante Ignacio Ramirez, Guillermo Prieto, Juan De Dios Peza, Jose Marti, Justo Sierra, Manuela Acuna and Rosario De La Pena.

Manuel Acuna:

I would like to clarify that is a placer see every body here.

Rosario De La Pena:

Of course all drunk.

One lovely drink like tequila and have the honor of our friend Rosario De La Pena.

Manuel como va el ultimo libro de poemas que estas escribiendo.

Manuel did you finished the last poetry book?

Besides being a student doctor from University National Autonoma from Mexico, where do you used to live?

Manuel Acuna:

I am kind of late finishing a book, but I have to go out from the religious Santa Briguida school for finishing my school of medicine in the University National Autonoma

Guillermo Prieto:

What kind of poems do you write?

Manuel Acuna:

I write romantic poems, kind of beautiful beens or humans with their own beauty.

Guillermo Prieto:

Angeles o que?

Angels or what?

Manuel Acuna:

Women that looks the way to declare love, but there is something that not let you speak with them, specially when you love some one but that person does not love you.

Guillermo Prieto:

The advise I give you is that there are a lot of women that wants to share their live with you dude.

Rosario De La Pena:

I would like to clarify that Manuel Acuna write beautiful poems.

Juan De Dios Peza:

Manuel I believe that your poems some kind elegant for Rosarito...maybe you like her...

Manuel Acuna:

Lets talk about something else please.

Rosario De La Pena:

What you can offer to a woman , life, pure soul, chastity, love...well tell me guys.

Guillermo Prieto:

I love those questions.

Juan De Dios Peza:

Mi love, tenderness, I take care of her like a jewel , of course with out showing her to somebody

Rosario De La Pena:

But no for that will be your slave ...buddy

Guillermo Prieto:

Of course not, but everybody knows that many guys like you, Rosarito you are so beautiful

Rosario De La Pena:

 We need more music.

 (In that moment Guillermo Prieto stand up from the table and tell the music guys play certain song.)

Guillermo Prieto:

Would you please sing last song for my beautiful friend Rosarito.

..Musico con guitarra:

(a guy playing and singing with a guitar)

 Claro que si...

Of Course

The music guy stand up and start singing

We can see the Manuel Acuna's face with a lot of passion while he hears the music and start crying. Manuel could give anything for Rosarito to kiss her or huge her with a lot of love.

Justo Sierra:

 I would like to clarify , thanks guys for invite me here, I would like to say thanks to Don Juan De La Pena and Mrs Margarita Llerena for invite me to their house and be in this great party.

Manuel Acuna:

I would like to dance with you mrs Rosario De La Pena…. Would you please give me that honor?

Rosario De La Pena:

Sera un placer ..

Will be a placer

(We can see Manuel Acuna how grab the hand of Rosario De La Pena and stars dancing with the music in that amazing saloon .

Manuel Acuna:

Why you dress in black

Rosario De La Pena:

I dress in black for one lover that passed awar.

Manuel Acuna:

Who is that lucky love ….that I can see through your eyes with a lot of passion

Rosario De La Pena:

I supposed marry Juan Espinoza De Los Monteros..

(We can see Rosario De La Pena looking to the sky like remembering the love that she lost …then there is a flash back and we can see Rosario De La Pena in the forest with Juan Espinoza De Los Monteros.

2-Ext/open field Verano/dia

We can see Juan Espinoza De Los Monteros kissing Rosario De La Pena they lay down on the grass forest and Juan Espinoza touch her legs so gently and finishing in love .

3- Int /Casa de la Familia de Rosario De La Pena hoy El Palacio de Bellas Artes./atardecer

We can see Manuel Acuna and Rosario de la pena dancing,

Manuel Acuna:

What happened later with your love?

Rosarion De La Pena:

There was an honor fight between lovers.

(vemos como hay un fade in /out)

3-Ext/ campo abierto/ al mediodía

There is a duel fight by guns , we can see Juan Espinoza De Los Monteros grabing one gun and the other guy another , they will fight for one love Rosario De La Pena… there is one bullet only in each gun

Testigo de Muerte:

There is a witness both signed a paper after taking a gun.. In that paper they signed that they will play to death Duelo de Muerte means some one will survived after this duel of guns.

Testigo De Muerte:

 Empieza a cntar 1….2….3….4….5…6….

The witness started count 1….2….3….4…5….6

(ambos empiezan a caminar opuestamente) y

Both started walking in opposite ways with the guns up…

Testigo De Muerte;

 7….8……9….fuego..

Witness of dead says

7..8..9....fire

We can hear the noisy bullets of the guns

And Juan Espinoa fell dead… we can see the face of Rosario De La Pena sad…and started yelling …why…why!

4-- Int /Casa de la Familia de Rosario De La Pena hoy El Palacio de Bellas Artes./atardecer

We can see Rosario De La Pena and Manuel Acuna dancing slow…. Rosario started crying and Manuel huge her with love.

Rosario De La Pena get out from the party and stop dancing

5- Int/ Universidad Nacional Autónoma de medicina de México./mediodía

Is the year 1868 and Manuel Acuna is taking a medicine class about the heart… We can see the doctor teacher talking about the leper decease ,

Maestro de Medicina:

We can see other decease like Galls.. and a news paper from New York one theory how to cure

Galls and other deceases like leper … we can see legs drawings so big and scratch, some blood on the leg like eruptions on the skin of the leg. There is certain cure with oregan, and 2 bottles of bitriolo substance helps disappear the pain with oxygen water.

Manuel Acuna:

 Could be possible this cure as well is good for the leper

Maestro De Medicina:

 This is a good question,,,now there is no cure for leper but is a possibility to cure leper in this way..

The leper decease is so old like the human kind, in the 14th century the leper people must make noisy with a bell to let know to the people that they have the decease .

(otro estudiante se levanta y pregunta)

Estudiante de Medicina:

Well the cure works on the legs from the horses we can put some Oregano and vitriola and the legs from the horse is cured and the leper or llaga disappear..

Maestro de Medicina:

In certain way the cure helps the horses but sometimes the human patients as well.

4.int/ Habitacion de Manuel Acuna/noche

Is dark to dark to see and we can see Manuel Acuna inside his bedroom writing a letter to Rosario De la Pena .

Manuel Acuna started writing a letter of love to Rosario De La Pena

Nocturno a Rosario

Manuel Acuna writes the letter with the light of a Oil candle .

MANUEL ACUÑA

Pues bien, yo necesito decirte que te adoro,

Well I have to tell you that I adore you,

Decirte que te quiero con todo el corazón;

Telling you I love you with all my heart

Que es mucho lo que sufro, que es mucho lo que lloro,

That is to much how much I suffer and how much I cry,

Que ya no puedo tanto, y al grito en que te imploro,

That I can not go aheat and I scream for you

Te imploro y hablo en nombre de última ilusión.

···I implore you in the name of the last illusion.

Comprendo que tus besos jamás han de ser míos,

I understand that your kisses never will be mine,

Comprendo que en tus ojos no me he de ver jamás;

I understand that through your eyes won't be seen myself

Y te amo y en mis locos y ardientes desvaríos,

And I love you, with my crazy though to love you

Bendigo tus desdenes, adoro tus desvíos,

I bless your love deny and adore your decisions.

 Y en vez de amarte menos te quiero mucho más.

So instead of loving you, less I love you …I love much more

…

Que hermoso hubiera sido vivir bajo aquél techo,

How beautiful could be live together under that ceiling

Los dos unidos siempre y amándonos los dos;

You and me together ,loving each others;

Tu siempre enamorada, yo siempre satisfecho,

You always in love, and I always satisfied

Los dos una sola alma, los dos un solo pecho,

Just two of us one soul, both of us with just one breast

Y en medio de nosotros mi madre como un Dios

And between us my mother like a God

…

Esa era mi esperanza…mas ya que a sus fulgores

That was my hope more than my anxiety

Se opone el hondo abismo que existe entre los dos,

The abyss is our enemy between us

¡Adiós por la última, amor de mis amores;

Goodbye love, of my love

La luz de mis tinieblas, la esencia de mis flores;

The darkness of my light, the essence of my flowers

Mi lira de poeta, mi juventud, adiós!

My poet writings , my youths of goodbye

(Al dia siguiente en los diarios de mexico mejor dicho periodicos aparece un encabezado donde se lee...

Manuel Acuna lo encuentran muerto intoxicado con alcohol y cianuro una carta en su mano derecha lo que parece una carta de amor a Rosario De La Pena.

The Next day we can see in the newspapers mentioned that Manuel Acuna killed himself with enough alcohol and poison cianuro potassium and in his right hand we found a love letter to his love Rosario De La Pena... Is as sad day today we say goodbye a poet, doctor and a lover Mexican Manuel Acuna

LETRAS DE DIOSES MEXICANOS

Written by Erik De La Torre Stahl copyrights 1990 Spanish version

En esta novela escrita por Erik De La Torre Stahl te invita a conocer a los grandes escritores y politicos mexicanos Manuel Acuna, Guillermo Prieto, Ignacio Ramirez. Una gran adagio novelistico tragico entre el escritor y poeta Manuel Acuna y Rosario De La Pena.

In this novel written by Erik De La Stahl invite you to know some of the amazing writers and poets from Mexico in the 1870's. This is the story of an amazing poet named Manuel Acuna in love with Rosario De La Pena

1. Int /Casa de la Familia de Rosario De La Pena hoy El Palacio de Bellas Artes./atardecer

Is the year 1873.

Is the year 1873 and is almost sunrise in the house Rosario De La Pena today Theatre Palacio de Bellas Artes.

Una gran tertulia en la Casa Paterna de Santa Isabel" La casa de la familia de Rosario De La Pena 26 anos, hija de Don Juan De La Pena y la sra. Margarita Llerena. (actualmente Palacio de Bellas Artes en la Cd. de México)

There is a great party in the ancestors family from Rosario De La Pena she is 26 years old her father Juan De La Pena and mrs Margarita llerena .

Allí se reunen el educador Gabino Barreda, El Nigromante Ignacio Ramírez, Guillermo Prieto, Juan de Dios Peza, José Martí, Justo Sierra, Manuel Acuña y Rosario de la Peña

We can see as well the teacher Gabino Barreda, El Nigromante Ignacio Ramirez, Guillermo Prieto, Juan De Dios Peza, Jose Marti, Justo Sierra, Manuela Acuna and Rosario De La Pena.

Manuel Acuna:

Quiero aclarar que es un placer verlos de nuevo.

I would like to clarify that is a placer see every body here.

Rosario De La Pena:

Especialmente verlos ebrios .

Of course all drunk.

El Nigromante:

 Una bebida espiritosa ,no hay como el tequila, estimado y ser los invitados de honor por parte de una hermosa dama como Rosario de la Pena..

One lovely drink like tequila and have the honor of our friend Rosario De La Pena.

Manuel como va el ultimo libro de poemas que estas escribiendo.

Manuel did you finished the last poetry book?

Aparte de ser estudiante de medicina de la Universidad Nacional Autonoma de Mexico. Donde vivíaz antes?

Besides being a student doctor from University National Autonoma from Mexico, where do you used to live?

Manuel Acuna:

 Me he dilatado en terminar el ultimo libro porque me tuve que salir del exconbento de Santa Briguida... y también tengo que atender mis estudios de medicina en la Universidad Nacional Autonomo de Mexico

I am kind of late finishing a book, but I have to go out from the religious Santa Briguida school for finishing my school of medicine in the University National Autonoma

Guillermo Prieto:

 Que tipo de poemas escribes Manuel

What kind of poems do you write?

Manuel Acuna:

Escribo poemas romanticos..seres hermosos incanzables..

I write romantic poems, kind of beautiful beens or humans with their own beauty.

Guillermo Prieto:

Angeles o que?

Angels or what?

Manuel Acuna:

Mujeres que buscas la forma de declarar tu amor a ellas, pero hay algo que te impide a hablar con ellas..

Especialmente cuando sabes que esa mujer que quisieras amar no te ama.

Women that looks the way to declare love, but there is something that not let you speak with them, specially when you love some one but that person does not love you.

Guillermo Prieto:

El consejo que te doy...hay muchas mujeres disponibles para compartir su vida con un hombre.

The advise I give you is that there are a lot of women that wants to share their live with you dude.

Rosario De La Pena:

Quiero aclarar que Manuel Acuna escribe muy bonitos poemas

I would like to clarify that Manuel Acuna write beautiful poems.

Juan De Dios Peza:

Manuel creo que tus poemas son elagantes para Rosarita...no será....

Manuel I believe that your poems some kind elegant for Rosarito...maybe you like her...

Manuel Acuna:

Cambiemos de tema por favor

Lets talk about something else please.

Rosario De La Pena:

Que le ofrecería una mujer a cambio de dar su vida, su castidad y amor por vivir con ustedes.

What you can offer to a woman , life, pure soul, chastity, love…well tell me guys.

Guillermo Prieto:

Me encantan esas preguntitas..

I love those questions.

Juan De Dios Peza:

Mi protección ,amor, cariño…la cuidaría como una joya sin ensenarla a nadie

Mi love, tenderness, I take care of her like a jewel , of course with out showing her to somebody

Rosario De La Pena:

Pero no por eso va ser tu esclava..amiguito.

But no for that will be your slave …buddy

Guillermo Prieto:

Pues claro que no Rosarito…al menos todos sabemos que muchos andan tras de tus huecitos hermosos.

Of course not, but everybody knows that many guys like you, Rosarito you are so beautiful

Rosario De La Pena:

Necesitamos mas música.

We need more music.

(En ese momento Guillermo Prieto se levanta de la mesa y va con el músico y le susurra en la oreja…)

(In that moment Guillermo Prieto stand up from the table and tell the music guys play certain song.)

Guillermo Prieto:

Podrías cantar la ultima para mi Dulcenea Rosarito….

Would you please sing last song for my beautiful friend Rosarito.

..Musico con guitarra:

(a guy playing and singing with a guitar)

Claro que si…

Of Course

(Se levanta el músico y empieza a cantar)

The music guy stand up and start singing

(nos percatamos que el rostro de Manuel Acuna se llena de pasión, al escuchar la música…y empieza a derramarse una lagrima.. Manuel daría cualquier cosa por tener en sus brazos a Rosario De La Pena)

We can see the Manuel Acuna's face with a lot of passion while he hears the music and start crying. Manuel could give anything for Rosarito to kiss her or huge her with a lot of love.

Justo Sierra:

Quiero agregar que es un placer estar aquí con ustedes amigos, y gracias principalmente a Don Juan De La Pena y la sra. Margarita Llerena por invitarnos a su casa y celebrar una tertulia mexicana.

I would like to clarify , thanks guys for invite me here, I would like to say thanks to Don Juan De La Pena and Mrs Margarita Llerena for invite me to their house and be in this great party.

Manuel Acuna:

Me gustaría bailara con usted señorita Rosario De La Pena….Me podría conceder tal honor

I would like to dance with you mrs Rosario De La Pena…. Would you please give me that honor?

Rosario De La Pena:

Sera un placer ..

Will be a placer

(vemos como Manuel Acuna toma de la mano a la señorita Rosario De La Pena, y empiezan a bailar al ritmo de la música que se escucha en el gran salón.)

(We can see Manuel Acuna how grab the hand of Rosario De La Pena and stars dancing with the music in that amazing saloon .

Manuel Acuna:

A que se debe que usted señorita se vista de negro..

Why you dress in black

Rosario De La Pena:

Guardo luto por un amor que se fue..

I dress in black for one lover that passed awar.

Manuel Acuna:

Quien fue ese amor afortunado...que la veo muy emocionada recordar tal suceso

Who is that lucky love ....that I can see through your eyes with a lot of passion

Rosario De La Pena:

Estaba comprometida con Juan Espinoza de Los Monteros.

I supposed marry Juan Espinoza De Los Monteros..

(nos damos cuenta que Rosario De La Pena se queda vehemente mirando al cielo pensando en su amado que perdió) y aparece una toma del campo en verano..y vemos Rosario De La Pena con Juan Espinoza De Los Monteros besándola sobre el campo..

(We can see Rosario De La Pena looking to the sky like remembering the love that she lost ...then there is a flash back and we can see Rosario De La Pena in the forest with Juan Espinoza De Los Monteros.

2-Ext/open field Verano/dia

Vemos como Juan Espinoza De Los Monteros besa con passion a Rosarion De La Pena..estan acostados sobre el campo abierto...rodeado de flores...y empieza a llover y empieza a besarse con pasión ....Rosario De La Pena sus labios tiemblan ...Juan Espinoza le levanta el vestido a Rosarion De La Pena y empiezan y le toca suavemente sus piernas tan hermosas y acaban en una comunión de cuerpo del amor....

We can see Juan Espinoza De Los Monteros kissing ?... De La Pena they lay down on the grass forest and Juan Espinoza touch her legs so gently and finishing in love .

Al final ellos se abrazan y lloran con la emoción que se sucito.

3- Int /Casa de la Familia de Rosario De La Pena hoy El Palacio de Bellas Artes./atardecer

Vemos como Manuel Acuna sigue bailando con Rosario De La Pena..abrazados uno al otro..y en ese momento le pregunta Manuel Acuna que paso después

We can see Manuel Acuna and Rosario de la pena dancing,

Manuel Acuna:

 Que paso después con tu amor

What happened later with your love?

Rosarion De La Pena:

 Se sucito un enfrentamiento de honor

There was an honor fight between lovers.

(vemos como hay un fade in /out)

3-Ext/ campo abierto/ al mediodía

Y nos percatamos que el amado de Rosario , Juan Espinoza De Los Monteros esta en duelo a muerte con una persona...ambos se encuentran frente a un testigo con una caja donde hay 2 armas de fuego..Juan Espinoza agarra una y la otra por el hombre desconocido...

There is a duel fight by guns , we can see Juan Espinoza De Los Monteros grabing one gun and the other guy another , they will fight for one love Rosario De La Pena... there is one bullet only in each gun

Testigo de Muerte:

 Ambos están de acuerdo y firman en esta acta que están en un acto de duelo de honor

(vemos como ambos firman acta de duelo de honor)

There is a witness both signed a paper after taking a gun.. In that paper they signed that they will play to death Duelo de Muerte means some one will survived after this duel of guns.

Testigo De Muerte:

 Empieza a cntar 1....2....3....4....5...6....

The witness started count 1....2....3....4...5....6

(ambos empiezan a caminar opuestamente) y

Both started walking in opposite ways with the guns up...

Testigo De Muerte;

 7....8......9....fuego..

Witness of dead says

7..8..9....fire

(Se escuchan disparos)

We can hear the noisy bullets of the guns

Y cae muerto Juan Espinoza.. vemos el rostro de Rosario De La Pena gritando...porque......porque!...

And Juan Espinoa fell dead... we can see the face of Rosario De La Pena sad...and started yelling ...why...why!

4-- Int /Casa de la Familia de Rosario De La Pena hoy El Palacio de Bellas Artes./atardecer

Vemos como Rosario De La Pena y Manuel Acuna bailan mas despacio...Rosario empieza a llorar y Manuel Acuna la abraza ..

We can see Rosario De La Pena and Manuel Acuna dancing slow.... Rosario started crying and Manuel huge her with love.

Rosario De La Pena Se retira de la tertulia...Y deja de bailar..

Rosario De La Pena get out from the party and stop dancing

5- Int/ Universidad Nacional Autónoma de medicina de México./mediodía

Es el ano 1868 y Manuel Acuna esta tomando una clase de medicina sobre el corazón. Vemos al maestro de medicina hablando sobre la cura de llagas en el cuerpo..inchazones por exceso de acido urico en el cuerpo

Is the year 1868 and Manuel Acuna is taking a medicine class about the heart... We can see the doctor teacher talking about the leper decease ,

Maestro de Medicina:

 En esta copia impresa por el periódico Estado Unidense en New York ..se presenta la teoría aprobada De la cura de Galls por el doctor C.L. Hammond que en ingles se significa cura de llagas o agallas en el cuerpo como razgunos, inchazones de la mano o extremidades del cuerpo como los pies..por gravedad y decantación los solidos en la sangre se acientan en los pies.. La clave y cura de esta enfermedad es 1 onza de oregano y 2 onzas de Vitriolo.... Se unta..y la inchazon desaparece en 3 dias.

We can see other decease like Galls.. and a news paper from New York one theory how to cure

Galls and other deceases like leper ... we can see legs drawings so big and scratch, some blood on the leg like eruptions on the skin of the leg. There is certain cure with oregan, and 2 bottles of bitriolo substance helps disappear the pain with oxygen water.

Manuel Acuna:

 Con respecto a la enfermedad Lepra...esta formula es curable para la misma

Could be possible this cure as well is good for the leper

Maestro De Medicina:

 Es una pregunta muy interesante hasta hoy no se ha descubierto la cura de la lepra, bueno creo como ustedes estudiantes y nosotros como doctores, tenemos una gran tarea poner a prueba toda enferedad que se presente...

This is a good question,,,now there is no cure for leper but is a possibility to cure leper in this way..

La lepra es tan antigua como la humanidad... en el siglo 14 Los leprosos tenían que sonar una campana para alertar a la gente de su enfermedad.

The leper decease is so old like the human kind, in the 14th century the leper people must make noisy with a bell to let know to the people that they have the decease .

(otro estudiante se levanta y pregunta)

Estudiante de Medicina:

 Las pruebas que se han hecho en caballos, para la cura de las llagas o razgunos han sido exitosas con este nuevo ungüento de Oregano y vitriola

Well the cure works on the legs from the horses we can put some Oregano and vitriola and the legs from the horse is cured and the leper or llaga disappear..

Maestro de Medicina:

 Ha sido exitosa la cura de llagas de caballos con este ungüento... también es cura para seres humanos..

In certain way the cure helps the horses but sometimes the human patients as well.

4.int/ Habitacion de Manuel Acuna/noche

Is dark to dark to see and we can see Manuel Acuna inside his bedroom writing a letter to Rosario De la Pena .

Vemos Manuel Acuna escribiendo una carta de amor a su amiga Rosario De La Pena.

Manuel Acuna started writing a letter of love to Rosario De La Pena

Y se escucha su voz en voz baja..escribe frente a una lámpara de aceite.

Nocturno a Rosario

Manuel Acuna writes the letter with the light of a Oil candle .

MANUEL ACUÑA

Pues bien, yo necesito decirte que te adoro,

Well I have to tell you that I adore you,

Decirte que te quiero con todo el corazón;

Telling you I love you with all my heart

Que es mucho lo que sufro, que es mucho lo que lloro,

That is to much how much I suffer and how much I cry,

Que ya no puedo tanto, y al grito en que te imploro,

That I can not go aheat and I scream for you

Te imploro y hablo en nombre de última ilusión.

···I implore you in the name of the last illusion.

Comprendo que tus besos jamás han de ser míos,

I understand that your kisses never will be mine,

Comprendo que en tus ojos no me he de ver jamás;

I understand that through your eyes won't be seen myself

Y te amo y en mis locos y ardientes desvaríos,

And I love you, with my crazy though to love you

Bendigo tus desdenes, adoro tus desvíos,

I bless your love deny and adore your decisions.

Y en vez de amarte menos te quiero mucho más.

So instead of loving you, less I love you ...I love much more

...

Que hermoso hubiera sido vivir bajo aquél techo,

How beautiful could be live together under that ceiling

Los dos unidos siempre y amándonos los dos;

You and me together ,loving each others;

Tu siempre enamorada, yo siempre satisfecho,

You always in love, and I always satisfied

Los dos una sola alma, los dos un solo pecho,

Just two of us one soul, both of us with just one breast

Y en medio de nosotros mi madre como un Dios

And between us my mother like a God

…

Esa era mi esperanza…mas ya que a sus fulgores

That was my hope more than my anxiety

Se opone el hondo abismo que existe entre los dos,

The abyss is our enemy between us

¡Adiós por la última, amor de mis amores;

Goodbye love, of my love

La luz de mis tinieblas, la esencia de mis flores;

The darkness of my light, the essence of my flowers

Mi lira de poeta, mi juventud, adiós!

My poet writings , my youths of goodbye

(Al dia siguiente en los diarios de mexico mejor dicho periodicos aparece un encabezado donde se lee...

Manuel Acuna lo encuentran muerto intoxicado con alcohol y cianuro una carta en su mano derecha lo que parece una carta de amor a Rosario De La Pena.

The Next day we can see in the newspapers mentioned that Manuel Acuna killed himself with enough alcohol and poison cianuro potassium and in his right hand we found a love letter to his love Rosario De La Pena... Is as sad day today we say goodbye a poet, doctor and a lover Mexican Manuel Acuna